MAD COWS
AND CANNIBALS

A Guide to the
Transmissible Spongiform Encephalopathies

CHARLOTTE A. SPENCER

University of Alberta

Pearson Education, Inc.
Upper Saddle River, NJ 07458

Editor-in-Chief, Life and Geosciences: Sheri L. Snavely
Executive Editor: Gary Carlson
Project Manager: Crissy Dudonis
Vice President of Production & Manufacturing: David W. Riccardi
Executive Managing Editor: Kathleen Schiaparelli
Assistant Managing Editor: Becca Richter
Production Editor: Elizabeth Klug
Supplement Cover Management/Design: Paul Gourhan
Manufacturing Buyer: Ilene Kahn
Electronic Composition and Formatting: William Johnson
Cover Photo Credit: John Redman/AP Wide World Photo

© 2004 Pearson Education, Inc.
Pearson Prentice Hall
Pearson Education, Inc.
Upper Saddle River, NJ 07458

All rights reserved. No part of this book may be reproduced in any form or by any means, without permission in writing from the publisher.

Pearson Prentice Hall® is a trademark of Pearson Education, Inc.

The author and publisher of this book have used their best efforts in preparing this book. These efforts include the development, research, and testing of the theories and programs to determine their effectiveness. The author and publisher make no warranty of any kind, expressed or implied, with regard to these programs or the documentation contained in this book. The author and publisher shall not be liable in any event for incidental or consequential damages in connection with, or arising out of, the furnishing, performance, or use of these programs.

Printed in the United States of America

10 9 8 7 6 5 4 3

ISBN 0-13-142339-8

Pearson Education Ltd., *London*
Pearson Education Australia Pty. Ltd., *Sydney*
Pearson Education Singapore, Pte. Ltd.
Pearson Education North Asia Ltd., *Hong Kong*
Pearson Education Canada, Inc., *Toronto*
Pearson Educación de Mexico, S.A. de C.V.
Pearson Education—Japan, *Tokyo*
Pearson Education Malaysia, Pte. Ltd.
Pearson Education, *Upper Saddle River, New Jersey*

Table of Contents

Preface

The transmissible spongiform encephalopathies (TSEs) may be the most bizarre and mysterious diseases in nature. They are unlike any other infectious diseases, are not caused by viruses or bacteria, and they challenge the accepted dogmas of molecular biology. These fatal neurodegenerative diseases are terrifying, appear to strike in a random manner, and have no cure. Most scientists now accept the idea that TSEs are caused by rogue protein molecules that convert normal proteins into abnormal ones; however, this concept is still controversial.

In the mid-1990s, the spread of Mad Cow Disease to humans triggered panic in Europe, crippled Britain's beef industry, and helped topple the Conservative government in the United Kingdom. In 2003, the discovery of a case of Mad Cow Disease in a single cow in Alberta, Canada led to bans on the export of Canadian beef, and raised questions about the extent of Mad Cow Disease in North America. In addition, concerns about the safety of the blood supply, vaccines, and surgical instruments keep Mad Cow Disease in the news around the world. In North America, an epidemic of chronic wasting disease currently threatens wild and domesticated deer and elk populations and has depressed some economies that rely on hunting. So little is known about TSEs that reliable screening tests and effective treatments still do not exist.

To comprehend the threat that TSEs pose to human health and to assess whether regulatory measures are sufficient to control their spread, we need to understand as much as possible about these diseases. The goals of this booklet are:

- to describe the nature of TSEs — their biology, history, and epidemiology;
- to address a number of specific questions about TSEs and human health with an emphasis on two recently emerged diseases — Mad Cow Disease and Chronic Wasting Disease;
- to pose some of the ethical questions that surround TSEs;
- to provide resources for further study.

We have attempted to present the most current and accurate information possible, derived primarily from scientific literature, the media, and government sources. As research is progressing rapidly and TSE epidemics are changing daily, readers are encouraged to refer to scientific journals and the Web sites listed in the References and Resources section for the latest developments.

Introduction

"But cows are herbivores, they shouldn't be eating other cows. It has just stopped me cold from eating another burger!"

Oprah Winfrey, transcript of The Oprah Winfrey Show, April 15, 1996.

Since the early 1900s, the Fore people in the East Highlands of Papua New Guinea honored their dead relatives through ritualistic cannibalism. Out of respect, they cooked and consumed nearly every part of the deceased's body. Women, infants, and young people ate most of the body: brain, internal organs, and bones that were baked, ground into a powder, and mixed with vegetables. The men, less involved in the funeral rituals, consumed muscle meat. The women and children smeared their bodies with tissue, which dried and remained on their skin for weeks. In this way, the Fore ensured that their dead would live forever, as part of their living relatives. These, too, would later be eaten by the tribe. Cannibalism of deceased relatives continued until 1960, when Western missionaries and the Australian Territorial government stopped cannibalism, converting these Stone Age people to cultivators of coffee beans.

In the 1920s, a mysterious and horrifying disease appeared among the Fore. It incapacitated and killed people of all ages, but primarily children and adult females. In contrast, adult males rarely contracted the disease. The first symptoms of this new disease were lack of coordination, staggering, and slurred speech. After a few months, victims would shiver uncontrollably and experience dramatic mood changes — from euphoria to indifference. They would then lapse into paralysis and die within a few months to a year. The Fore people called the disease "kuru," meaning "to tremble with fear." They believed that it was caused by evil sorcery. When some victims of kuru realized their fate, they would try to deny their condition. Others would simply withdraw from the tribe, refuse food, and die of suicidal starvation.

By the 1950s, kuru had become a serious epidemic. More than 2,500 people died between 1957 and 1975, threatening the entire population of the Fore (about 30,000). The nature of kuru baffled scientists. It was not a psychological disease, it was not caused by a conventional virus or bacterium, and it was not a genetic condition. Nor was it due to heavy metal or other poisoning, or a vitamin deficiency. People with kuru developed neurological symptoms and their brains were full of sponge-like holes and abnormal deposits of protein. What seemed clear, though, was that kuru was linked to cannibalism. The incidence of kuru declined gradually after cannibalism was banned. Within 4 years, it disappeared from the 4- to 9-year-olds, and within 10 years it declined in the 10- to 19-year-olds. Children born after the cessation of cannibalism never developed the disease. However, adults came down with kuru well into the 1990s, showing that kuru has a potentially long incubation time (40 years). Another baffling feature was that the infective agent of kuru survived the cooking processes that the Fore used to prepare bodies for mortuary feasts.

What was the nature of this mysterious disease? The answer to this question began to appear in the 1960s, but even today it is still not completely resolved. We now know that kuru is a progressive degenerative brain disease spread by eating the most infectious parts of the body — brain and nervous tissue. For the Fore people, the disease passed to women and children who ate infected brains. When they died of kuru, they would be eaten by their kinsmen as well, spreading the disease in a tragic, escalating cycle (Fig. 1).

Figure 1
A Fore man, holding the body of a young kuru victim.

Forty years after the cessation of cannibalism ended the cycle of kuru in New Guinea, another strange new disease appeared half way around the world. And another form of cannibalism — this time among cattle — was likely the cause.

Between 1984 and 1986, dairy cows across England began to develop strange neurological symptoms. They became apprehensive or aggressive, had muscle tremors, lost weight, and were uncoordinated. The disease was inevitably fatal after a few months to a year. Laboratory diagnosis showed that the brains of these cows were full of sponge-like holes and protein deposits, similar to those found in brains of kuru victims. This condition, called bovine spongiform encephalopathy (BSE) or "Mad Cow Disease," was declared a new cattle disease in 1986 (Fig. 2). Over the next 6 years, BSE raged into an epidemic, with 180,000 confirmed cases by the year 2000. Unfortunately, the number of confirmed cases is surpassed by estimates of unconfirmed cases. It is estimated that about 2 million animals were infected with BSE but were asymptomatic, and of these, about 1.6 million would have entered the human food chain. Humans likely ate millions of burgers and tons of frozen processed meat containing BSE-infected material before the epidemic subsided and government regulations reduced the amount of infected tissue used as food. It became evident that BSE had a long incubation time of at least 5 years, meaning that many cattle were silently incubating the disease and were probably entering the human food chain from at least the early 1980s.

Scientists suspected that agricultural feeding practices were spreading BSE. At the time, cattle were fed cheap high-protein supplements to enhance growth and production. These supplements, called "meat and bone meal," were prepared by cooking, mashing, and drying down the carcasses of animals including the brains of sheep, goats, and cattle — a process called *rendering*. The remains of cattle that were nonproductive or showing signs of illness were sent to rendering plants, becoming food for other cattle. Like cannibalism and kuru in humans, bovine cannibalism accelerated the cycle of BSE infectivity. In 1988, the British government banned the use of cows and sheep in feed for other cows and sheep. Today, BSE is subsiding as an epidemic, although it continues to be detected in British cattle and is now appearing in Europe, Japan, and the Middle East.

Figure 2
The British government slaughtered and incinerated over 4 million British cattle in an attempt to control the BSE epidemic.

Although the BSE epidemic was tragic enough for farmers and the British meat industry, the full impact was not felt until 1996, when the British government announced that Mad Cow Disease had infected humans. They acknowledged that ten young patients had been diagnosed with unusual neurological symptoms — uncoordinated movement, shaking, and psychiatric problems. These patients degenerated rapidly and died within a few months to a year. Their brains resembled those of kuru victims and BSE-infected cattle — full of sponge-like holes and protein deposits. Evidence was compelling that this new disease in humans was caused by eating BSE-infected beef (Fig. 3).

The political and economic fallout was immediate and intense. The European Union invoked a worldwide ban on British beef and beef products, leading to the near-bankruptcy of the $6 billion beef industry. The government was forced to slaughter over 4.6 million British cattle, creating hundreds of thousands of tons of carcasses that will take until the year 2004 to incinerate. It cost over $12 billion to compensate farmers, import milk, and restock cattle herds. The way in which the British government had handled the BSE epidemic contributed to the

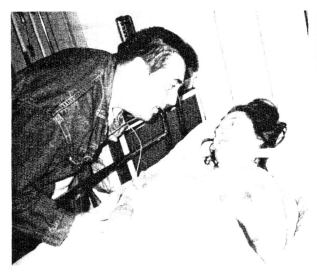

Figure 3
Michelle Bowen, 29, was one of the first victims of the human form of Mad Cow Disease. Seen here with her husband, Tony, Michelle died in Manchester, England in November 1995.

defeat of the Conservative government in 1997. Mad Cow Disease triggered public panic and was one of the worst health crises of the twentieth century.

To date, more than 120 people, mostly in the United Kingdom and Europe, have died from the human form of Mad Cow Disease, and the incidence is increasing by about 20% each year. No one knows how many people will succumb to the disease over the next few decades. Estimates range from a few hundred to 150,000.

What are these strange and frightening brain diseases? What causes them and how are they spread? Can they be treated or cured? Can we prevent new diseases like BSE from infecting humans?

In this guide, we will attempt to answer these questions and present some of the fascinating stories that surround the transmissible spongiform encephalopathies — the TSEs.

The Biology of Transmissible Spongiform Encephalopathies

WHAT ARE TSEs?

Kuru, BSE, and the human form of BSE are all members of a group of neurological degenerative diseases known as transmissible spongiform encephalopathies (TSEs). These diseases affect humans and animals, and are invariably fatal. The TSEs have long incubation periods and have characteristic pathological features: the brain is riddled with holes resembling those in a sponge (hence, "spongiform") and contains protein deposits called *plaques*. Unlike bacterial or viral infections, there is no inflammatory response to TSEs, and TSE victims do not make anti-TSE antibodies (molecules produced by the immune system that recognize and destroy infectious and foreign substances). The TSE victims lose motor function, become demented, and eventually die. Human TSEs include kuru, Creutzfeldt-Jakob disease (CJD), Gerstmann-Sträussler-Scheinker syndrome (GSS), fatal familial insomnia (FFI), and new variant CJD (nvCJD) — the human form of Mad Cow Disease. The TSEs occur in many animals as well, including cattle, sheep, goats, mink, deer, elk, cats, and exotic ungulates (Table 1). Scrapie in sheep is the best-studied of the animal TSEs and has been endemic in sheep flocks for centuries.

THE HUMAN TSEs

Creutzfeldt-Jakob Disease

Creutzfeldt-Jakob disease (CJD) was first identified in the 1920s, and is the most common TSE in humans. It occurs worldwide at the rate of 1 person per 1 million per year, and strikes people predominantly between the ages of 55 to 75. Patients show a rapid progressive dementia, visual problems, speech abnormalities, and muscle incoordination. They also display tremors and behavioral problems such as agitation and depression. Death occurs within 12 months for most patients. The brains of CJD patients characteristically show widespread spongiform changes and moderate levels of protein deposits.

The disease occurs in several forms: sporadic, familial, and iatrogenic. The sporadic form accounts for 85% of cases, appears to arise spontaneously, and is not associated with any risk factor that has been studied (such as diet, environment, or genetics). Familial CJD accounts for about 10% of cases. It is associated with specific dominant mutations in the prion protein gene (*Prnp*) — which will be discussed in more detail below. The transmissibility of CJD was first noted in 1968, when CJD was transferred by intracerebral inoculation of brain tissue from a CJD patient into a chimpanzee. Transmission from person to person occurs in iatrogenic (or "caused by medical treatment") CJD. This form of CJD has occurred through corneal or dura mater (brain lining) grafts, through the use of contaminated surgical instruments, or by injection of growth hormone derived from human pituitary glands. Over 90 people worldwide have

Table 1 The Transmissible Spongiform Encephalopathies in Humans and Animals

Abbreviation	Disease	Host	Mechanism of Infection
-	Kuru	humans	Ritualistic cannibalism
sCJD	sporadic Creutzfeldt-Jakob Disease	humans	Mutation or spontaneous conversion of PrP^c to PrP^{sc}
iCJD	iatrogenic Creutzfeldt-Jakob Disease	humans	Tissue grafts, surgical instruments, human growth hormone
fCJD	familial Creutzfeldt-Jakob Disease	humans	Heritable mutations in the *Prnp* gene
nvCJD	new variant Creutzfeldt-Jakob Disease	humans	Ingestion of BSE-infected meat
GSS	Gerstmann-Sträussler-Scheinker Disease	humans	Heritable mutations in the *Prnp* gene
FFI	Fatal Familial Insomnia	humans	Heritable mutations in the *Prnp* gene
-	Scrapie	sheep	Ingestion or contact with prion-infected secretions
BSE	Bovine Spongiform Encephalopathy	cattle	Ingestion of contaminated feed
TME	Transmissible Mink Encephalopathy	mink	Ingestion of contaminated feed
FSE	Feline Spongiform Encephalopathy	cats	Ingestion of BSE-contaminated feed
-	Exotic Ungulate Encephalopathy	kudu, oryx, nyala	Ingestion of BSE-contaminated feed
CWD	Chronic Wasting Disease	deer, elk	Contact with, or ingestion of, prions

developed CJD after injection with growth hormone purified from cadavers. The incubation times for CJD range from 3 to over 20 years. Even though growth hormone is no longer collected from human pituitary glands, but is synthesized by recombinant DNA technology, more cases may emerge owing to the long incubation times for CJD.

Gerstmann-Sträussler-Scheinker Syndrome and Fatal Familial Insomnia

Both GSS and FFI are inherited diseases, associated with the presence of dominant mutations in the *Prnp* gene. The GSS syndrome is considered a variant of CJD, with similar clinical features to CJD: motor disorders, dementia, and the characteristic spongiform encephalopathy. A rare disease, GSS occurs at the rate of 5 cases per 100 million. Fatal familial insomnia shows a different clinical pattern to CJD or GSS, with progressive insomnia as a predominant feature. Although protein deposits are found in FFI patients' brains, no spongiform changes occur.

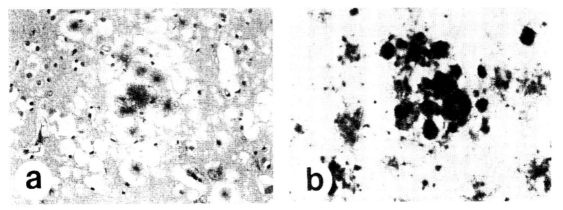

Figure 4

Appearance of brain sections from nvCJD patients. (A) Section from frontal cortex, showing protein plaques, surrounded by spongiform degeneration. (B) Plaques that stain positive for PrP protein.

New-variant CJD

New-variant CJD (nvCJD) first appeared in 1996, as a form of CJD with unique properties that distinguish it from classic CJD. Patients are unusually young (16 to 39 years old) compared to classic CJD patients (over 50 years old). Patients first show behavioral symptoms such as aggression, anxiety, apathy, depression, paranoid delusions, or withdrawal. Only after about 6 months do they show neurological symptoms such as shaking, incontinence, and immobility. Their brains resemble those of kuru patients (or scrapie-infected sheep), rather than those of classic CJD patients. Although spongiform changes occur throughout the brain, nvCJD patients have unusually large deposits of protein, especially in cerebral and cerebellar regions — a feature not found in classic CJD. The plaques are described as "florid" in appearance, surrounded by halos of spongiform holes (Fig. 4). These distinctive clinical features, combined with epidemiological evidence (such as appearance of nvCJD in the United Kingdom only during and after the BSE epidemic), provided the initial link between nvCJD and BSE.

The link between BSE and nvCJD was strengthened by laboratory studies. For example, researchers in France inoculated monkeys with brain material from BSE-infected cattle. Within 4 months, these monkeys showed the same behavioral and neurological symptoms as seen in nvCJD patients. In addition, the monkeys' brains contained the same type and patterns of spongiform changes and protein deposits seen in nvCJD patients, but not in classical CJD patients.[1] The strongest evidence that nvCJD is the human version of Mad Cow Disease came from a study using mice. Scientists injected brain homogenates from patients with sporadic CJD or nvCJD, or homogenates from BSE-infected cattle, into the brains of mice. Mice injected with nvCJD or BSE material showed almost identical disease patterns, in terms of incubation times, disease progression, symptoms, and appearance of spongiform and plaque deposits in the brain. Mice injected with sporadic CJD material showed very different patterns, with longer incubation times, different symptoms, and distinct brain

---

[1]Lasmezas et al., Nature 381: 743 (1996).

changes.[2] In addition, analysis of proteins in the brains of nvCJD patients showed that they resembled proteins in BSE-infected brains but not those in CJD-infected brains.[3] These data provided clear evidence that the same infectious agent was responsible for both BSE in cattle and nvCJD in humans.

WHAT CAUSES TSEs?

For many years, TSEs resisted analysis. The diseases are difficult to study, as they require injection of infected brain material into the brains of experimental animals, and the disease takes months to years to develop. In humans, TSEs are not transmitted by normal contact, but require the infectious material to be introduced into the body by ingestion or injection. The TSE victims do not develop antibodies to the disease agent. At the present time, the only way to make a firm diagnosis of a TSE is to examine brain tissue after death. Initially, scientists thought that TSEs were caused by viruses or even smaller entities called "viroids" that act with long incubation times. However, a virus could never be isolated. The infectious agent is not affected by high doses of ultraviolet or ionizing radiation or by nucleases that damage nucleic acids, bacteria, or viruses; however, it can be partially destroyed by some reagents that hydrolyze or modify proteins.

In the early 1980s, American scientist Stanley Prusiner purified the infectious agent from scrapie-infected sheep brain, and concluded that it consists only of protein. He proposed that scrapie is spread by an infectious protein particle that he called a *prion*. "Prion Hypothesis" was dismissed by many scientists. The idea that an infectious agent could self-replicate, but contain no genetic material was, a heretical one. (Box 1). However, Prusiner and others have presented evidence supporting the Prion Hypothesis, and the notion that diseases can be transmitted by particles that contain no DNA or RNA, although still controversial, has gained acceptance.[4]

HOW CAN INFECTIOUS PROTEINS CAUSE DISEASE?

The Prion Hypothesis

The protein-only Prion Hypothesis has the following features:

- *Normal nerve cells contain the normal prion protein.*

   Normal cells in the brain and other tissues synthesize a glycoprotein called PrP^c (prion protein cellular), encoded by the *Prnp* gene. [A glycoprotein is a protein molecule that is covalently linked to one or more carbohydrate molecules.] The PrP^c protein is found predominantly on the outer membrane of neurons and some other cell types. The PrP^c protein can be readily digested by proteases (enzymes that degrade proteins) and is denatured and destroyed by heat, detergents, and chemicals. Its function is still unknown.

- *TSE-infected nerve cells contain the abnormal form of the prion protein.*

   The active component of prions is an abnormal version of the PrP^c protein called PrP^{sc}

[2]Bruce et al., Nature 389: 498 (1997).
[3]Hill et al., Nature 389: 448 (1997).
[4]Prusiner, Proc. Natl. Acad. Sci. USA 95: 13363 (1998).

(prion protein scrapie). The PrP^sc protein has unusual properties. It is insoluble, and its infectivity survives high temperatures, radiation, detergents, disinfectants such as alcohol, formalin, and phenol and hence is almost indestructible. The PrP^sc protein is not fully degraded by protease enzymes. Both PrP^c and PrP^sc differ from one another only in their three-dimensional structures. The PrP^c protein is composed mostly of coils termed α-helices, whereas PrP^sc contains a predominance of flat β-sheets, over α-helices, (Fig. 5). The PrP^sc protein is found inside affected cells, rather than on the cell surface.

Prion protein scrapie accumulates in the brains of TSE-infected animals but is absent in healthy control animals, suggesting that the abnormal protein contributes to the development of the disease. The way in which PrP^sc may lead to the destruction of brain cells is still unclear.

- *PrP^sc interacts with PrP^c, converting PrP^c into another PrP^sc.*

The initiating event in the development of TSEs is the conversion of a normal PrP^c protein to the abnormal PrP^sc form. When the normal PrP^c protein contacts a prion PrP^sc molecule, the normal protein is somehow unfolded and refolded into the abnormal conformation. This conformational change may result from PrP^sc acting as a template that directs the conversion of PrP^c to PrP^sc. Once the normal PrP^c molecule has been transformed into the abnormal form, it spreads its lethal conformation to other normal PrP^c molecules. The process then takes off in a chain reaction, filling the brain with the insoluble PrP^sc protein.

- *Infectious PrP^sc can come from inside or outside the host.*

The source of abnormal PrP^sc can be either external (from ingestion or injection of prion material from another organism), or internal (from a spontaneous change in the conformation of a normal PrP^c molecule to the PrP^sc form). The spontaneous internal change may be simply a random protein-misfolding event (leading to development of sporadic CJD), or may be due to the presence of mutations in the *Prnp* gene that make the mutated PrP^c protein more susceptible to spontaneous refolding into the PrP^sc form (leading to development of familial forms of CJD).

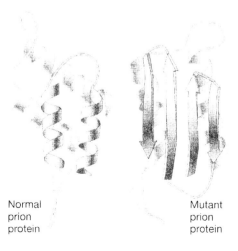

Figure 5
The normal protein, PrP^c (left), contains a predominance of α-helices. The abnormally folded prion protein, PrP^sc (right), contains a high percentage of β-sheets.

Normal
prion
protein

Mutant
prion
protein

Box 1 The Central Dogma

The Central Dogma of Molecular Biology states that genetic information flows in one direction — from DNA to RNA to protein. First, the information contained within genes (in the form of DNA sequence) is decoded into RNA molecules through the process of transcription. Second, RNA molecules are decoded into proteins through the process of translation.

There are some exceptions to the Central Dogma. First, information can flow backwards from RNA into DNA by reverse transcription — a process that occurs in some viruses. Second, both DNA and RNA can be replicated into copies of themselves. However, there has yet to be a serious challenge to the concept that proteins are the end state and ultimate goal of genetic information. It is believed that proteins can only be created using templates provided by DNA and RNA — i.e., they are not self-replicating.

Proteins make up the structure of our cells and carry out the enzymatic functions that allow us to live. The unique combination of proteins within a cell determines its phenotype or function. Proteins are composed of subunits called *amino acids*, linked together in a linear fashion, like links on a chain. There are 20 different types of amino acids, and their order in a protein (known as the *primary sequence*) determines the identity of the protein. For example, the human β-globin protein contains 147 amino acids, in a specific order. The prion protein contains 245 amino acids. After a protein chain is synthesized, the linear chain of amino acids folds into a specific shape, or conformation. This is known as the protein's secondary or tertiary structure. The conformation of a protein determines how it will function. (continued on next page)

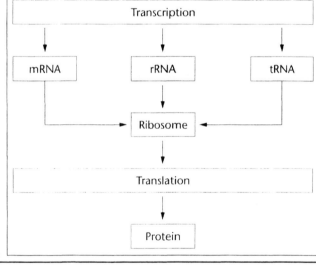

Figure A
Simplied view of the Central Dogma: information flows from DNA to RNA to protein.

(continued on next page)

Evidence to Support the Protein-Only Prion Hypothesis

Considerable evidence now supports the Prion Hypothesis. For example:

- Experimental mice that lack the *Prnp* gene, and hence make no PrP^c protein, cannot be infected with prions and do not accumulate PrP^{sc} in their brains. If the *Prnp* gene is reintroduced into these mice, they synthesize PrP^c protein, and subsequently they become susceptible to prion infection.

- If mice that lack the *Prnp* gene are grafted with brain tissue from mice that have the gene, the grafted tissue develops a TSE pathology, but the surrounding tissue does not.

- In the test tube, purified PrP^{sc} can bind to purified PrP^c, converting it into an insoluble form indistinguishable from PrP^{sc}.

Box 1 (continued)

Different amino acid sequences dictate different conformations and hence different functions.

In summary, the sequence of nucleotides in DNA codes for the sequence of ribonucleotides in RNA. The sequence of ribonucleotides in RNA codes for the sequence of amino acids in proteins. The sequence of amino acids in proteins dictates the protein's conformation and hence its function. Mutations in a gene alter the sequence of nucleotides in the DNA, and may ultimately alter a protein's amino acid sequence, conformation, and function.

Many scientists consider the protein-only Prion Hypothesis to be heretical, because it proposes replication of, and information transfer between, proteins — in the absence of DNA or RNA. However, the Prion Hypothesis may also be considered consistent with the Central Dogma, as the PrP^c protein is synthesized in cells by the normal processes of transcription (from the *Prnp* gene) and translation. The conversion of PrP^c to PrP^{sc} may be thought of as a chemical conversion, rather than true replication of a disease agent.

Regardless of how one defines the nature of the prion agent, it is one of the most bizarre and puzzling entities in the biological world.

(a) Primary structure

– Ala – Glu – Val – Thr – Asp – Pro – Gly –

(b) Secondary structure

α helix

β sheet

(c) Tertiary structure

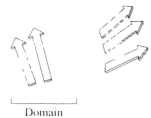

Domain

Figure B
Levels of protein structure. (a) Primary Structure: a linear sequence of amino acids. (b) Secondary Structure: conformations of the linear chain. (c) Tertiary Structure: the fully folded chain.

- Mice that contain a *Prnp* gene bearing the same mutation as found in human patients with GSS spontaneously develop a TSE resembling Gerstmann-Sträussler-Scheinker Syndrome (GSS). This "GSS" TSE can be transmitted by injection to other mice that do not bear that particular mutation, illustrating both the inheritability and transmissibility of TSEs.

The decisive experiment that would test the protein-only Prion Hypothesis would be to synthesize PrP^c in a test tube and convert it into PrP^{sc} in the absence of any nucleic acids or small viruses that might contaminate proteins purified from living cells. If this synthetic PrP^{sc} protein caused a spongiform encephalopathy in an experimental animal, the protein-only Prion Hypothesis would be difficult to dispute. To date, such an experiment has not been possible, as the PrP^{sc} protein has been difficult to synthesize.

Other Important Features of TSEs and Prions

Prion Strains

Prions seem to have distinct strains that "breed true" through serial infections. For example, prions from scrapie-infected sheep can be divided into groups that differ in their incubation times, patterns of brain defects that they cause, as well as the pattern of carbohydrate molecules linked to the PrP^{sc} protein. Proponents of the protein-only Prion Hypothesis explain this strain-specificity by invoking the "conformational hypothesis." This states that each prion strain, rather than having a distinct amino acid sequence, simply has a unique structure or conformation. Each of these conformational strains can convert the host's PrP^c proteins into copies of its unique strain conformation.

The Prion Species Barrier

An important feature of TSEs is that they are generally transmitted inefficiently between species, and when they do cross the species barrier, their incubation times are very long. For example, scrapie in sheep has never been observed to infect humans, despite its prevalence in sheep populations for centuries. It is also difficult to infect mice with scrapie prions, even if they are injected directly into the mouse brain.

The prion species barrier may be due to differences between the amino acid sequences of each species' PrP^c protein, as encoded in their *Prnp* genes. The species barrier may also be affected by the conformational differences that exist between the PrP^c proteins from each species, in part as a result of the amino acid sequence differences. For example, hamster prions usually do not cause disease in mice. Hamster and mouse *Prnp* genes encode PrP^c proteins that differ from each other by 10 amino acids out of a total of 245. However, if mice are engineered to bear copies of the hamster *Prnp* gene (as well as their own *Prnp* genes), they can make both mouse and hamster PrP^c proteins. When injected with mouse prions, they accumulate mouse PrP^{sc} in their brains; when injected with hamster prions, they accumulate hamster PrP^{sc}. In other words, prions more efficiently interact with, and alter the conformation of, PrP^c molecules that match them in amino acid sequence.

Leaping the Species Barrier

A disturbing feature of prions is that their infectivity can be modified after passage through another species, resulting in a less effective species barrier. When prions from one species are introduced into another, the TSE incubation time is long. If the host's new prions are then transferred to another host of the same species, the incubation time is considerably shortened. For example, mice that are fed BSE prions develop the disease, whereas hamsters do not. However, if hamsters ingest brain homogenates from BSE-infected mice, they develop the infection. The passage of BSE through another species (mouse) produces prions that are competent for infection in a previously resistant species (hamster).

Although prions are usually transmitted inefficiently between species, the story of BSE and nvCJD appears to be an exception. There are over 30 amino acid differences between bovine and human PrP^c proteins; however, the species barrier was readily crossed. This may indicate that other factors beyond amino acid sequence contribute to the species barrier, such as aspects of protein conformation or interaction of prion proteins with other host factors. It is also possible that specific regions of the PrP^c protein are more important than others for determining the species barrier, and amino acid differences in these small regions are more important than overall sequence differences between the prion proteins. Brain homogenates from BSE-infected cattle cause disease after injection into cattle, sheep, goats, mice, pigs, marmosets, and mink, but not hamsters or chickens. Also, BSE has infected cats and zoo animals through contaminated feed.

Mutations and Inherited TSEs

More than 20 different mutations in the human *Prnp* gene have been linked to inherited TSEs. These mutations alter one or more amino acids in the sequence of the PrP^c protein. According to the Prion Hypothesis, these amino acid changes may make the mutant PrP^c protein more susceptible to spontaneous refolding from PrP^c to PrP^{sc}. The length of time required for a spontaneous refolding event (or an accumulation of unfolding events) to take place may explain why inherited forms of TSEs take years to manifest themselves.

Genetic Variation and TSE Susceptibility

Animals within a population differ in their susceptibility to TSEs, depending upon variations (polymorphisms) in the amino acid sequence of their normal PrP^c protein. For example, the human *Prnp* gene encodes polymorphisms at amino acid position 129 of the protein. Over 90% of iatrogenic or sporadic CJD patients have *Prnp* genes that code for only methionine or only valine at this position. However, 50% of the general population has this genetic trait. As there have been only about 120 deaths from nvCJD to date, and about 18 million British people have *Prnp* genes that code for only methionine or valine at position 129 of the PrP protein, other factors beyond this genetic trait must be involved in determining susceptibility to TSEs. It is still not known how genotype contributes to susceptibility to TSEs.

Questions About TSEs and Their Transmission

HOW DO PRIONS GET INTO THE BRAIN?

To study TSEs, scientists inject tissue homogenates into the brains of experimental hosts. Although this is an efficient method of transmitting TSEs, in real life TSEs are transmitted by eating infected material (kuru and nvCJD), by intravenous or intramuscular injection, through tissue transplantation, or from contaminated surgical instruments (iatrogenic CJD). To develop therapies to TSEs, and to understand which tissues contain infectious prions, it is important to understand how infectious prions move from the site of infection to the central nervous system where they trigger the formation of protein plaques and destroy neurons.

Prions appear to pass through the intestinal wall rapidly, then enter lymphoid tissues such as lymph nodes and spleen where they are thought to incubate and perhaps replicate. A functional immune system is required for prions to enter the central nervous system, as well as fully differentiated B cells (cells that produce antibodies for the immune system). Prions are thought to be taken up by peripheral nerves and then transported to the spinal cord, and the brain. After sheep are fed BSE-containing material, prion protein can be detected in the digestive tract, lymph nodes, peripheral nerves, spinal cord, and brain. The possibility that lymphoid tissues and B cells of the immune system may harbor prions has raised questions about the infectivity of blood from BSE-infected animals, including humans (see below, for discussion of blood transfusions and TSEs). In addition, it may suggest that animals infected with prions, but showing no symptoms, may sequester prion proteins in other tissues besides spinal cord, and brain. In patients with nvCJD, prion proteins are detectable in tonsils and appendices, as well as neural tissue.

DO SURGICAL AND DENTAL INSTRUMENTS SPREAD TSEs?

The possibility that TSEs such as nvCJD could be transmitted through the use of contaminated surgical or dental instruments is also of concern. Prions can adhere to metal surfaces and are difficult to inactivate by standard sterilization measures. Prions are highly resistant to radiation, heat, and chemical disinfectants. They are not destroyed by boiling or normal cooking processes. Even standard autoclave sterilization (steam heat at 121°C for 15 minutes) does not fully inactivate them. They survive treatment with disinfectants such as alcohol, formaldehyde, and glutaraldehyde, and are resistant to strong alkali treatment for 1 hour. One study showed that exposure of scrapie prions to dry heat temperatures of at least 600°C was required for complete inactivation. Prions can even survive in soil for up to 3 years (Box 2). To inactivate prions, they must be treated for 1 hour with a bleach solution containing at least 2% chlorine. In addition, a combination of alkali and pressure heating (autoclaving) at 132°C for 1 hour kills prion infectivity. Autoclave sterilization at 134°C for 18 minutes can reduce prion infectivity a thousand-fold.

Box 2

"These agents are almost immortal. They resist alcohol, they resist boiling, they resist hospital detergents. We thought it would be interesting to see what would happen if we buried some of these agents, and so I ground up some scrapie brain and mixed it with soil, put it in a flower pot, enclosed it in a cage, and used my own garden as a burial site — right here. And what we found was that a good deal of the infectivity remained in the soil after three years. We exposed it to temperatures that turned it to ash, and it did not entirely kill the agent, and so every known pathogen of man would have been destroyed by this process, and this was not."

Dr. Paul Brown

Laboratory of Central Nervous System Studies, National Institutes of Health,
Bethesda, MD 20892, USA

From the PBS Nova program "The Brain Eater," February 10, 1998

The extreme resistance of prion proteins to inactivation by normal decontamination procedures makes them difficult to eradicate from food, blood, or surgical instruments. Hospitals in the United Kingdom are required to clean all surgical instruments thoroughly before sterilization so as to reduce the amount of organic material that might harbor prion proteins. If surgery is performed on people suspected of having, or known to have, CJD or nvCJD, instruments must either be disposed of by incineration, or quarantined awaiting the patient's diagnosis. As prions cannot be realistically inactivated by standard decontamination procedures, these measures can only minimize the risk of transmission.

The U.S. Centers for Disease Control also recommends that stringent decontamination procedures be used on surgical instruments that contact high infectivity tissues (such as brain, spinal cord, and eyes) and lower infectivity tissues (such as cerebrospinal fluid, liver, lymph nodes and spleen) from patients that are either known to have, or suspected of having, some form of CJD. These same decontamination methods are recommended after surgery on people who are genetically related to those with inherited forms of TSEs. These stringent measures include careful washing, and decontamination in either strong alkali or high chlorine bleach plus autoclaving (steam heat at 121°C for 30 minutes). Surfaces and heat-sensitive instruments must be washed in strong alkali or bleach for at least an hour. Routine decontamination methods are recommended for instruments that contact tissues of people not suspected of being at risk for a TSE.

There are six known cases of iatrogenic CJD being spread by surgical instruments and EEG (electroencephalograph) electrodes. As yet, there have been no confirmed cases of nvCJD being transmitted by reused surgical instruments. It is not known whether nvCJD prions are present in sufficient quantities in tissues other than neural tissues to present a risk for transmission to patients through contaminated surgical instruments after surgery not involving brain or spinal cord.

There is no evidence that dental procedures increase the risk of contracting TSEs. However, animal experiments have shown some infectivity in gingival tissues and dental pulp. Prions have also been detected in nerves of the mouth and face. In the United Kingdom, it is recommended that patients with TSEs, or at risk for developing these diseases, undergo dental treatment in hospital settings. The World Health Organization recommends that general infection-control methods during and after dental procedures are sufficient when treating TSE patients, as long as neurovascular tissue is not involved.

CAN TSEs BE SPREAD THROUGH BLOOD TRANSFUSIONS OR BLOOD PRODUCTS?

In 2002, the U.S. Food and Drug Administration (FDA) introduced stringent new regulations governing who can donate blood. These regulations ban donors who have resided in the United Kingdom for 3 months or more between 1980 and 1996, donors who have lived in Europe for 5 years or more between 1980 and present, and donors who have received a blood transfusion in the United Kingdom between 1980 and the present. These restrictions have not been welcomed by blood bank officials, who question the necessity for such strict rules, given the lack of evidence that CJD or nvCJD can be transmitted through the blood supply. The new regulations are particularly unwelcome in metropolitan New York, which has relied on Europe for 25% of its blood supply.

In the United Kingdom, blood donor regulations are even more strict. All donated blood has the white cells (including B cells) removed, as some evidence suggests that prions may be associated with these cells. In addition, the United Kingdom imports blood plasma from the United States for transfusions into children born after 1996, to protect them from a theoretical risk of nvCJD exposure. Plasma from the United States is also imported for the manufacture of blood products such as clotting factors. The British government is considering banning blood donations from donors who have ever received a blood transfusion, a move that may create a 10% decline in blood donations.

Are these strict regulations necessary? There has never been a documented case of iatrogenic CJD or nvCJD that was acquired through blood transfusion or blood products. In contrast, experimental evidence suggests the possibility of blood infectivity. Blood from animals experimentally infected with scrapie or CJD is infectious after introduction into the brain (intracerebral injection) or into the abdominal cavity (intraperitoneal injection) of another animal. Also, nvCJD prions have been detected in lymphoid tissues, suggesting a theoretical possibility of blood transmission. Given that nvCJD is such a new disease with an unknown incubation period, it may take years or decades to detect infections that have occurred through blood transfusion.

Until 2002, officials considered that the risk of being infected by blood donated by a patient with nvCJD was low and "theoretical." This assessment was upgraded to "an appreciable risk" following publication of a study of BSE-infected sheep.[5] In that study, sheep were experimentally infected with BSE, then used as blood donors to healthy sheep. Two of 24 sheep developed BSE after blood transfusion and another two sheep were showing symptoms at the time the study was published. The two BSE cases had been transfused with whole blood from sheep that were still healthy and were incubating BSE asymptomatically. In a parallel study, four out of 21 sheep transfused with blood from scrapie-infected sheep came down with scrapie, also demonstrating the transmissibility of TSE prions in blood. The results of this study would seem to indicate that the regulations to safeguard blood supplies from TSE prions may be justified, especially considering that the number of people who may be incubating nvCJD, but not showing symptoms, is unknown.

WHAT ABOUT ORGAN TRANSPLANTS?

Sporadic or familial CJD can be transmitted through organ transplants, including corneal and dura mater (lining of the brain) grafts, as well as by injection of pituitary-derived human growth hormone. About 200 patients worldwide have contracted iatrogenic CJD by these methods. As

Hunter et al., J. Gen. Virol. 83: 2897 (2002).

yet, no case of iatrogenic transmission of nvCJD has been documented. However, given the possible long incubation time of nvCJD and the fact that nvCJD is a relatively new disease, any iatrogenic transmissions may take years to be detected.

The FDA is presently considering whether to restrict residents from the United Kingdom from donating organs. Other countries such as Australia recommend that organ donors disclose the time they spent in the United Kingdom, but each organ or tissue transplant case is assessed on a case-by-case basis, weighing the potential risks against the benefit to the recipient. As there is no screening test yet for TSE infections, it is not possible to decrease any theoretical risk to zero, unless organ and blood donor programs cease.

ARE VACCINES SAFE?

Over the centuries, vaccines have been effective weapons in our fight against many serious infectious diseases. Vaccines are made from killed or weakened viruses or bacteria (or their component parts). Injection or ingestion of these vaccine components triggers a person's immune system into producing specific antibodies that recognize and destroy any invading virus or bacterium that the person later encounters.

In October 2000, Britain recalled a commonly used oral polio vaccine after discovering that it had been manufactured using fetal calf serum derived from British cattle. Since 1989, Britain had recommended that vaccine manufacturers not use bovine materials from BSE-affected countries and these same recommendations had been put forward by the European Union in 1999.

Since 1993, the FDA has recommended that vaccines not be manufactured using bovine-derived material from countries that have BSE. However, in February 2001, regulators found that five drug companies had manufactured vaccines using blood, serum, and meat broth derived from cattle in BSE-suspect countries. Nine different vaccines including polio, diphtheria, tetanus, and anthrax were in question. The FDA decided to allow these vaccines to be used, but to replace them over a 1-year period.

What are the risks posed by these vaccines? As yet, there are no known instances of humans contracting CJD or nvCJD from vaccines, nor is there any evidence that BSE prions are present in vaccines. To evaluate the answer to this question, it is necessary to understand how vaccines are manufactured. To prepare bacteria and viruses for vaccines, large quantities of these organisms must be grown in rich liquid nutrient media. Nutrient media that support the growth of bacteria often contain beef broths. Small quantities of bacteria are inoculated into the media in which they replicate to high numbers. The bacteria are then isolated from these growth media and components are processed for injection. Unlike bacteria, viruses grow only within other cells and are hence propagated in cell cultures. These cell cultures require the presence of calf serum (typically 10 to 20%) in media used for their growth. The viruses are isolated from these cultures and processed for injection or oral administration. If BSE prions are carried over to the end point of vaccine manufacture, what would be the risks of these vaccines being infective? The FDA estimates that the risk of contracting nvCJD from a vaccine prepared using BSE-infected material would be about 1 case in 40 million doses. They also emphasize that the benefits of vaccinating against serious infectious diseases such as polio and diphtheria far outweigh the unknown risks of contracting nvCJD.

ARE THERE DIAGNOSTIC TESTS FOR TSEs?

To date, there is no efficient way to detect TSEs in live patients at early stages of infection. Animals with TSEs do not launch an immune response to the PrP^{sc} protein, and therefore an antibody-based blood test for prion diseases is not possible. Detecting CJD or nvCJD in patients before they show clinical symptoms is important, as it could reduce the chance that they could transmit the disease through blood/organ donation or through contamination of surgical equipment. It could also allow early treatment of the disease, once a treatment has been discovered. Research is continuing on development of diagnostic tests. Some interesting possibilities are the detection of PrP proteins in urine and an electrocardiogram test that appears to detect heartbeat anomalies at the early stages of the disease. There are reports that a blood test for nvCJD may be available by 2004.

ARE THERE TREATMENTS FOR TSEs?

At the present time, the only treatment for patients with TSE diseases is palliative care. Over 60 potential drugs have been tested on experimental animals, and some are effective at reducing or eliminating the diseases if administered before or early in infection. These drugs do not work well, though, when administered after symptoms appear. As there are presently no tests that detect early TSE infection, drugs that act in early stages will not be beneficial to most patients at this time. However, early administration of drugs could be beneficial to people who have recently been exposed to infectious tissue grafts or contaminated surgical instruments. A few drugs have been tested in humans, with disappointing results. However, these have not yet been tested in large clinical trials, so their efficacy is still not clear. These drugs include antivirals such as acyclovir, the antimalarial drug quinacrine, and the anti-psychotic drug chlorpromazine.

The ideal treatment would be prevention of TSEs by vaccination. The fact that PrP^{sc} is simply a misfolded form of a normal cellular protein (PrP^c) makes it a less than ideal target for the immune system. However, research is continuing on devising an anti-TSE vaccine.

Questions About Mad Cow Disease and nvCJD

WHY DID BSE SUDDENLY APPEAR IN BRITISH CATTLE IN THE 1980s AND WHY DID IT BECOME AN EPIDEMIC?

First detected in British cattle between 1984 and 1985, BSE was declared a new disease in 1986. Its origins are still a mystery, but three predominant hypotheses exist. First, it is possible that BSE appeared in cattle after they became infected with scrapie prions that were present in cattle protein supplements. For decades, cows were fed cheap protein supplements to increase their growth and milk yield. These supplements were prepared in rendering plants and feed mills, by collecting the carcasses of animals that were unfit for human consumption, processing them by cooking and solvent extraction, and preparing a ground-up material called "meat and bone meal." In Britain, the rendered carcasses would certainly have contained scrapie-infected sheep, including their brains and nervous tissue. Scrapie has been endemic in sheep flocks for centuries, without causing detectable problems for humans or other animals.

In the late 1970s, rendering practices changed in Britain. The solvent extraction step, followed by pressurized steam treatment, was omitted. It is thought that this change in rendering procedure may have allowed scrapie prions to remain in the meat and bone meal, which was then eaten by cattle. As discussed earlier in this guide, prions can be modified by passage through a different species. This modification may make them more infective in another species. It is possible that scrapie prions were modified in cattle, making them more infective to other cattle. The BSE-infected cattle were then rendered into animal feed, escalating the epidemic in a cycle of cannibalism reminiscent of kuru in New Guinea. In 1988, the feeding of ruminant-derived protein to other ruminants (cattle, goats, sheep) was banned, and all cattle suspected of having BSE were slaughtered and disposed of — some simply by burial, others by incineration. The BSE epidemic has subsided since 1993 (Fig. 6); however, the disease has not disappeared.

The second hypothesis is that BSE first arose spontaneously within one British cow in the 1970s. This cow was rendered into meat and bone meal, spreading BSE to other cattle who in turn entered the bovine food chain. The third hypothesis is that BSE has been endemic in cattle for many years. It was only when feeding practices became cannibalistic and rendering practices were modified that BSE-infected material spread the disease as an epidemic throughout Britain.

Other possible contributors to the British BSE epidemic were delays in enacting regulations dealing with animal feed and lapses of both enforcement and compliance with regulations. For 2 years after BSE was recognized as a disease, BSE-infected cattle were rendered and fed back to other cattle, prior to the ban in July 1988. In addition, as the incubation time of BSE in cattle appears to be approximately 5 years, it is likely that some cattle were silently incubating BSE in the 1970s and 1980s, and they could have spread the disease when they were rendered into cattle feed. Even after the ban, it is estimated that only half of the prohibited ruminant material was properly separated and processed in rendering plants, even up until 1994. Farmers continued to feed infected supplements to cattle, so as to use up old stocks of feed, and because old feed

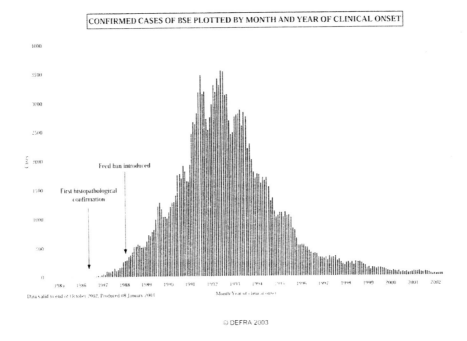

© DEFRA 2003

Figure 6
Confirmed cases of BSE in the United Kingdom, plotted by month and year of clinical onset, to 1999. (From the UK Ministry of Agriculture, Fisheries and Foods, 1999)

contaminated farm equipment. It is believed that BSE-contaminated feed was not completely disposed of until 1996, after the British government initiated a feed-recall scheme, followed by cleaning and disinfection of farm premises.

WHY IS BSE STILL PRESENT IN BRITISH CATTLE?

In 1988, the British government banned cattle and sheep remains being fed back to cattle and sheep, although feed containing cattle and sheep could be fed to poultry and pigs and feed containing poultry and pigs could be fed back to cattle. In 1994, the European Union (EU) prohibited the use of all mammalian protein in ruminant feed. And in 1996, all livestock were prevented from eating meat and bone meal prepared from cattle and sheep. Despite these tough measures, cattle born after the bans are still coming down with BSE.

Several possible scenarios could explain why British cattle are still being infected with BSE, even after the feed bans:

- First, it is possible that feed is still being contaminated with the remains of BSE infected cattle, due to noncompliance of renderers and feed mills, or from cross contamination of feed with old supplies of meat and bone meal.

- Second, it is possible that the BSE prion passes from animal to animal on pastures that are contaminated by the excrement of infected cows, or by contact with animals that are silent carriers. It is possible that BSE prions could persist for years in soil or water. Scrapie in sheep is known to be transmitted from animal to animal through contact with sheep feces or by ingestion of placental or fetal tissues. If it turns out that BSE can be spread horizontally by contact with infected environments, it may never be eradicated from British cattle. Early in the BSE epidemic, about 10,000 cattle were slaughtered and buried at sites throughout Britain. It is estimated that about 40 of these cattle may have been infected with BSE. The British government is now concerned about the potential for environmental contamination and is checking these burial sites to determine whether they represent a risk to animal or human health.

- Third, it is possible that there are as yet unrecognized sources of BSE infection. For example, a recent outbreak of BSE in Japan signals a potentially new route of transmission — through milk substitute. Five cases of BSE have appeared in Japan since 2001, and none of these animals were fed meat and bone meal. However, during the first month or two of life, all had been fed a milk substitute manufactured by the same company. This milk substitute contained sugars, skimmed milk, and animal fat (tallow). Some scientists suspect that tallow may be the source of BSE prions, and tallow has not been thoroughly studied as a source of TSE infection. The Japanese tallow had been imported from the Netherlands, where a BSE outbreak was occurring. If milk or tallow are confirmed as sources of BSE prions, it could have serious implications for present regulations over human and animal feed processing. At the moment, these substances are considered low risk for BSE spread, and they have been permitted in rendered animal feeds for ruminants.

Although any of these possible routes of infection may have occurred, the reasons for the post-ban BSE cases in Britain are still unknown, and research into their causes will take years to complete.

WHICH COUNTRIES HAVE BSE-INFECTED CATTLE, AND HOW DID THE DISEASE GET THERE?

To date, BSE has been detected in every European country except Sweden. Cases have also appeared in the Middle East, Japan and Canada. The latest estimates are that up to 2 million British cattle have been infected with BSE since the early 1980s. There are several possible explanations for how BSE spread from Britain to other parts of the world. First, some countries imported BSE-infected cattle from the United Kingdom prior to the implementation of their own import bans. This was the case in Canada, which imported a BSE-infected cow in 1987. After the infection was discovered in 1993, the infected cow and its 400 herd mates were slaughtered as a preventive measure. It is estimated that more than 50,000 cattle were exported from Britain after 1985, to countries around the world. Second, some nations may have imported British meat and bone meal. Over 70,000 tons of British meat and bone meal (containing remnants of cows, including neural tissue) were exported from Britain up to 1990. This material was intended for poultry and pig feed; however, given the ease by which feed can contaminate other feed, it is possible that BSE-containing protein supplements were fed to cattle after 1990.

IS IT SAFE TO EAT BEEF IN THE UNITED KINGDOM AND EUROPE?

This question is probably impossible to answer accurately and depends on having knowledge of BSE and nvCJD that is simply not available at the moment. For example, we would need to know:

How Many Cattle Are Infected With BSE?

Statistics show that several hundred cattle in Britain are diagnosed with BSE each year, and the disease has been detected in other countries as well (Fig. 7). Unanswered questions are whether a larger number of cases go undetected, and whether apparently healthy cattle that are incubating BSE can transmit infectious prions. Cattle over 30 months old are banned from human consumption, because the disease takes several years to develop. It is assumed that during the long incubation time (over 5 years), insufficient BSE prions are present in these cattle to transmit the disease. However, this remains an assumption.

Figure 7
Italian health workers inside barrier erected on a farm suspected of harboring a case of Mad Cow Disease.

Which Parts of the Animal Contain BSE Prions?

It has long been known that prions concentrate in nervous tissue — brain and spinal cord. However, other areas of the body may contain prions. When TSEs are transmitted through ingestion, prions first pass through the intestinal wall. They may then form reservoirs in the immune system (spleen, tonsils, lymph nodes) and may be associated with B cells (present in bone marrow and blood) before travelling through the central nervous system to the brain. As a precaution against the risk of BSE exposure, bovine brain, spinal cord, intestines, internal organs, and tonsils were banned from human consumption in 1989 in the United Kingdom. The practice of stripping meat from bovine backbone was banned in 1995. Meat on the bone was also banned, but the prohibiton was lifted in 1999. Milk, muscle meat, tongue, gelatin, and blood are presently considered low risk and fit for human consumption.

Some recent studies question these assumptions of low risk. In October 2002, new data indicated that bovine tonsils could transmit BSE to other cattle. Although tonsils are removed from cattle carcasses at slaughter, residues may remain on tongue tissue. The British government is considering extending the human consumption ban to bovine tongue. Another study in 2002 indicated that muscle tissue of TSE-infected mice can replicate prions and transfer the disease to other mice;[6] however, these data have been challenged by other studies that use different experimental methods. If muscle tissue from BSE-infected cattle is shown to be infectious, it challenges all the safeguards that have been used to protect consumers and increases the risks of nvCJD infection considerably. The subject of blood as a source of BSE prions is also controversial; however, new studies suggest that it, too, may harbor infection. The topic of transmission through blood is discussed earlier in this guide.

Are BSE Prions Entering the Food System and if so, in What Amounts?

Like the previous question, an accurate answer requires information that is currently unavailable. In 2000, it was estimated that up to 300 infected cattle may reach the food chain in the United Kingdom each year. Also, it is not known how many cattle at the early asymptomatic stages are being slaughtered for consumption, or whether these asymptomatic cases of BSE contain sufficient prions to pose a health risk to humans.

Even if a certain number of BSE-infected cattle are slaughtered, government regulations stipulate that highly infective parts of the animal are removed. Although this is likely to remove the majority of potentially infective material, mistakes and the inherent "sloppiness" in the slaughter process could lead to contamination of meat. For example, in September 2002, a European Commission inspection discovered that workers in British slaughterhouses were cross-contaminating carcasses with spinal cords during the slaughter process. By October 2002, UK inspectors had discovered over 13 cases of imported European beef contaminated with bovine spinal cord, even though EU laws require the removal of brain and spinal cord at slaughter.

Even if regulations are strictly adhered to in abattoirs, slaughter practices themselves could contribute to the spread of prions. The preferred method of stunning animals for slaughter is to use the "captive bolt pistol" to fire a retractable bullet into the front of the brain. Although unconscious, the animal's heart continues to beat for several minutes. Two recent studies show that this slaughter practice fragments the brain, pieces of which enter the blood system and the

---

[6]Bosque et al., Proc. Natl. Acad. Sci. USA 99: 3812 (2002).

animal's tissues. It has been shown that brain-associated material can be detected in the abattoir environment, on hands of workers, on slaughter equipment, and in samples of blood, internal organs, and muscles of slaughtered animals. It is not known how much this procedure may contribute to the spread of BSE prions or whether it increases the risk to humans.

What Constitutes an Infectious Dose of BSE Prions in Humans?

Ingestion of about 1 teaspoon of BSE-infected brain is sufficient to cause a BSE infection in cattle, and about ten-fold more than this is sufficient to cause an infection in mice. But no one knows what the infectious dose for humans is, or whether the dose can be accumulated over time. As a degree of species barrier exists between cattle and humans, the human lethal dose is likely to be more than the bovine lethal dose, but it is simply not known.

How Susceptible Are Individual Humans to the Disease?

As explained earlier in this guide, certain human *Prnp* genotypes may be particularly susceptible to nvCJD. However, many other factors may contribute to one's susceptibility, including environmental factors, co-infections with other agents, and gut permeability. There is no information on these aspects of nvCJD susceptibility at present.

What About Other Kinds of Meat — Lamb, Pork, Fish, Chicken?

Sheep and goats are susceptible to scrapie — a TSE that has been endemic in Britain for centuries and in the United States since 1947. About 500 scrapie cases are reported each year in Britain, although the real number may be much larger. There is no evidence that scrapie can be transmitted to humans. The real concern is that other animals such as sheep may harbor BSE prions asymptomatically, and transfer them to humans through food. Sheep can be experimentally infected with BSE by injection or ingestion of bovine BSE-infected brain material. Also, sheep were fed BSE-infected meat and bone meal in the 1980s at the same time that cattle were fed these same protein supplements and the BSE epidemic began. As BSE resembles scrapie, cases of sheep BSE may have gone undetected. Studies to determine whether BSE is present in sheep have not been completed as yet. If BSE is present in the United Kingdom sheep population, it could considerably increase the number of human deaths from nvCJD, over those due to bovine BSE alone. With stringent controls over feeding practices and human consumption, this number could be reduced. At the moment, EU rules prohibit consumption of brain, spinal cord, spleen, tonsils, and eyes from sheep more than a year old. The topic of BSE in sheep is currently of great concern in Britain, and may not be resolved for several years.

It is not known whether TSEs exist in pigs, poultry, or fish. Like other vertebrates, these animals have PrP^c proteins, so it is theoretically possible for them to develop TSEs. It is unknown whether these animals could be infected with BSE, or act as silent carriers. Until recently, all EU countries except the United Kingdom fed meat and bone meal (containing rendered cattle and sheep) to pigs, chicken and fish. Although this practice was banned in Britain in 1996, it is still permitted in the United States and Canada.

A disturbing report in July 2002 suggested that some chicken meat in the United Kingdom and the Netherlands had been adulterated with beef proteins, which could contain BSE prions. Chicken is sometimes injected with water and protein to make it absorb water and hence increase its weight — a process called "tumbling." One of the potential problems with this practice is that the source of protein, if not regulated, could be obtained from anywhere, including countries in which BSE is found. Apparently it is legal to inject chicken with beef and pork protein as long as the chicken is labeled as such. If this report is substantiated, it could lead to another food scandal in Europe and another source of BSE risk for humans. The process of tumbling also occurs in the United States; however, the type of protein added to the poultry must be indicated on the label. As import of beef products from countries with BSE is prohibited, it is unlikely that BSE-infected protein would be added to U.S. chicken.

In summary, no one can accurately assess the risks of developing nvCJD from eating beef (or other meat) from countries where BSE occurs. The risk is considered to be low, but probably not zero (Box 3). These questions will only be answered in the coming decades, as more information is obtained about TSEs and nvCJD.

Box 3 Traveler's Health Information — "Risk to Travelers"

The U.S. Centers for Disease Control (CDC) published a notice to travelers on the CDC Web site stating that the current risk of acquiring nvCJD from eating beef and beef products from European cattle cannot be determined precisely. However, the CDC estimates that the risk may be in the range of 1 case per 10 billion servings in the United Kingdom, and less in other European countries. Nonetheless, it advises travelers to avoid beef and beef products or to select only solid pieces of muscle meat. Milk and milk products are considered safe for consumption.

HOW MANY PEOPLE HAVE DIED FROM nvCJD AND HOW MANY WILL DEVELOP THE DISEASE?

As of December 2002, a total of 119 people in the United Kingdom have died of nvCJD, as reported by the CJD surveillance unit in Edinburgh. Another 10 people were diagnosed with the disease but were still alive. The United States reported its first case of nvCJD in 2002 in a 22-year-old woman who had been raised in England and had recently moved to the United States. A death due to nvCJD occurred in Canada in 2002, in a person who visited Britain frequently. Other cases have appeared in Italy (2), France (6), Ireland (1), and South Africa (1).

So little is known about nvCJD that making predictions about the course of the epidemic is extremely difficult. The incidence of nvCJD increased by about 20% per year between 1994 and 2001. Estimates of future total cases have ranged from as low as 100 to as high as 500,000. Predictions are based on the assumption that the BSE epidemic is under control, few humans will be newly infected, and that the incubation time is decades — assumptions that cannot yet be verified. Also, it is not known how susceptible the population is to the disease. A realistic estimate is thought to be less than 10,000 cases. If sheep are found to harbor BSE, the nvCJD estimates could increase considerably, to as many as 150,000. There is no reliable screening test for people incubating nvCJD, making estimates of future cases even more difficult. A recent survey of tonsils and appendices removed during operations in the United Kingdom detected BSE prions in one out of over 8,000 samples. Although the numbers are inadequate to accurately assess future nvCJD cases, these data suggest that about 120 people per million (about 7,000) may show signs of nvCJD in the future.

DOES MAD COW DISEASE OR nvCJD OCCUR IN NORTH AMERICA?

To date, only two cases of nvCJD have been reported in North America: one in the United States (a recent immigrant from the United Kingdom) and one in Canada (a frequent traveler to Britain).

Bovine spongiform encephalopathy has not been detected in U.S. cattle. It is known that 334 cattle were imported into the United States from Britain between 1980 and 1989. The fate of over half of these animals is not known. In 1989, the U.S. Department of Agriculture banned the import of live ruminants and most ruminant products from the United Kingdom and other countries with BSE. The ban was extended to all European nations in 1997. In Canada, one cow imported from the United Kingdom was diagnosed with BSE in 1993. This animal and its herd mates were destroyed and incinerated as a preventive measure. In 2003, a cow in northern Alberta tested positive for BSE. Although the animal was sick at the time of slaughter and did not enter the human food chain, it was rendered into feed for non-ruminants. Over 2,800 animals at risk for contracting BSE from this cow were slaughtered and non were found to have BSE. Several of the cow's herd mates had been imported into the US in 1997; however, it is not known whether these cattle were infected with BSE. Although it is likely that the Alberta cow contracted BSE from eating contaminated feed prior to 1997, it is known whether the feed was domestic or imported. Canada's import rules are similar to those in the United States.

A number of other regulations have been designed to prevent or contain an outbreak of BSE in the United States should it arise in American cattle or be introduced from outside the country. In 1997, the FDA banned the use of most (but not all) mammalian protein in animal feed intended for ruminants. A number of proteins are exempt from this rule, including blood, blood products, gelatin, milk, milk products, proteins derived from pigs or horses, and meat products derived from restaurant or institutional plate waste. It is also legal in the United States and Canada to feed rendered cattle carcasses to pigs, horses, and poultry, and to feed rendered carcasses of pigs, horses, and poultry to cattle, as long as the packaging is labeled appropriately. As mentioned previously, there is now some question about the infectivity status of blood and tallow, and it is not known whether nonruminant animals could act as carriers or reservoirs for BSE prions. Both U.S. and Canadian rules contrast with those of the EU, which prohibit the feeding of any mammalian protein back to other animals. In 2002, inspections of rendering and feed companies in the United States showed that 95% were compliant with feed regulations, an improvement over inspections in 2001, which showed only 78% of firms as compliant.

In 2000 and 2001, the U.S. government banned all imports of rendered animal proteins regardless of species, as well as animal feed from countries at risk for BSE. Exceptions are milk, milk products, hides and skins, protein-free tallow, and gelatin for pharmaceuticals. Between 1990 and 2000, the U.S. Department of Agriculture examined approximately 11,000 brains from cattle showing abnormal behaviors (out of about 300 million slaughtered during this period). No BSE was detected.

Following the discovery of a case of BSE in 2003, Canada introduced measures to exclude certain high risk materials from human food. These include bovine brain, spinal cord, eyes, tonsils, and part of the small intestine. To date, the US has not banned these materials from human consumption. At present, the United States and Canada do not exclude meat removed mechanically from bones, and meat derived from the vertebral column. It has been suggested that this regulation be introduced to prevent any BSE present in cattle from being transmitted to humans through food.

Other sources besides the food supply could introduce BSE or nvCJD into North America. These might include:

- imported drugs, blood products, vaccines, transplant tissues, dietary supplements, and cosmetics — or their raw materials — from BSE-infected sources.
- direct infection from humans infected with BSE in other parts of the world, then subsequently transmitting nvCJD iatrogenically.

The FDA and U.S. Department of Agriculture have little power over the manufacture of dietary supplements suspected of containing bovine tissues from unregulated sources. The FDA recommends that bovine products from countries with BSE not be used in dietary supplements; however, they cannot enforce this recommendation. Dietary supplements may contain glandular material from pituitaries, prostates, or kidneys, even brains in some cases. A recent report by a scientific advisory panel to the EU reported that tons of European beef products are shipped to the United States annually. These include gelatin, collagen, semen, and albumin. These are not used in food, but potentially are used in cosmetics or for vaccine manufacture.

Vaccine manufacturing has also been questioned recently. The FDA recommends that manufacturers of pharmaceutical products not use bovine-derived materials in the production of vaccines. However, some companies have not complied with these recommendations. In 2001, the FDA announced that several firms may have used bovine material from unknown geographical locations to make vaccines. These materials include serum and protein broths to grow cell cultures that support production of viruses.

A risk-assessment report released by Harvard University in 2001 concludes that U.S. measures are sufficient to prevent BSE from becoming a public health problem, even if it does arise within the cattle population. However, some critics of current regulations suggest that until extensive testing confirms the absence of BSE in North American herds, it will be only a best-guess whether the disease is present. The United States currently tests about 1 out of 18,000 cows slaughtered (2,300 out of 36 million slaughtered in 2000), compared with Switzerland, which tests about 1 out of every 60 cows (14,000 out of 800,000 slaughtered). Also, the United States uses a less-sensitive test than the one used in Europe. Some scientists believe that the current BSE testing program in the United States would be inadequate to detect a BSE incidence even as high as that in France (which detected 235 cases in 2002). Dr. John Collinge, a neurologist at University College in London and an expert in BSE, believes that the United States should test cattle herds more extensively. "Every country in Europe went through a phase of denying they had a problem," Dr. Collinge said. "After mandatory testing was introduced last year, countries that denied it vehemently discovered that they did have the disease." (Quoted in the *The New York Times*, Dec. 1, 2001)

Questions About Chronic Wasting Disease: The North American Mad Cow Disease?

The 2002 autumn deer hunt in Wisconsin was like no other. In the preceding 6 months, 24 white-tail deer in southern Wisconsin tested positive for chronic wasting disease (CWD), a transmissible spongiform encephalopathy that affects deer and elk. Although there is no evidence that humans have contracted the disease, the fears are spreading. A third of Wisconsin's hunters plan to avoid the hunt altogether, and others will not eat the meat. Some families are discarding frozen venison from previous years. It is estimated that the fears and uncertainties over CWD may severely reduce the $1 billion income that deer hunting brings to the state. State officials are enlisting hunters to eradicate 25,000 deer in an affected zone in southern Wisconsin, and they will test another 25,000 deer throughout the state in late 2002. The hunt has been extended into 2003 to enable this large kill. Officials will examine up to 50,000 deer brains, and dispose of about 2 million to 3 million pounds of deer by high-temperature incineration. The whole operation will cost over $2.5 million. The disease appears to be spreading through captive and wild populations of deer and elk in several U.S. states and two Canadian provinces. To release sufficient funds to deal aggressively with CWD, the U.S. Department of Agriculture issued a declaration of emergency because of CWD — "an emergency that may threaten the livestock industry in the U.S."

WHAT IS CWD?

Chronic wasting disease is a fatal neurological disease that affects wild and farmed members of the cervid family (deer and elk). It is a TSE disease that resembles scrapie, BSE, and human CJD. Affected animals become uncoordinated, lose weight, exhibit behavioral changes, and have increased salivation, drinking, and urination. Death occurs by emaciation or secondary pneumonia (Fig. 8). There is no known prevention, treatment, or efficient diagnostic test for live animals. Definitive diagnosis can be made only by examining brain tissue after death. Brains of CWD-infected animals show spongiform changes and protein plaques, as seen in other TSEs. The disease appears to have an incubation time of at least 2 to 3 years in deer and elk. As in other TSEs, CWD prions are detectable in brain, spinal cord, and lymph nodes.

WHERE DID CWD COME FROM?

The disease was first detected in 1967 in a mule deer in a wildlife research station in northern Colorado. The first cases of wild deer and elk with CWD appeared in the early 1980s, in northern Colorado and southern Wyoming. Since 1996, CWD has been detected in farmed elk herds in Colorado, Kansas, Montana, Nebraska, Oklahoma, South Dakota, Minnesota, and the Canadian provinces of Saskatchewan and Alberta. Recently, CWD appeared in wild deer in northwest Nebraska, southern New Mexico, southwest South Dakota, southern Wisconsin, northwest Colorado, and the province of Saskatchewan.

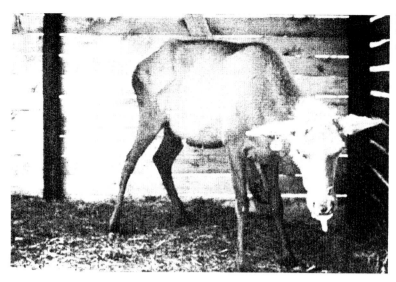

Figure 8
Captive elk with chronic wasting disease.

It is thought that the disease spread rapidly as a result of selling and transporting infected domestic deer and elk. The elk industry exploded in the 1990s, as elk velvet is valued in the Far East as a traditional medicine, bringing up to $200 a kilogram. Elk and deer meat are also prized as a source of low-fat protein in the domestic market. By 2001, about 450 elk had been shipped from Colorado to other states, including Wisconsin, which is now experiencing an epidemic of CWD in wild deer. Minnesota ranchers are thought to have trucked over 4,000 elk and deer from other states, some of which are now known to harbor cases of CWD.

The origin of CWD is still a mystery. It is thought that the disease may have been present for 35 to 40 years in wild populations in areas such as Colorado that now have concentrations of the disease estimated at 1 to 10% of the wild population.

HOW IS CWD RELATED TO BSE AND nvCJD?

As far as anyone knows at present, no direct connection exists between CWD and nvCJD besides their classification as TSEs. Although it has been conjectured that the first cases of CWD may have occurred after deer were fed cattle feed containing BSE prions, no direct evidence exists that TSE-infected feed was responsible.

HOW IS CWD SPREAD FROM ANIMAL TO ANIMAL?

One significant difference between BSE and CWD is that CWD appears to be spread from animal to animal by contact, and may not require ingestion of infected neural tissue. This contact may consist of touching or ingesting an infected animal's body fluids such as urine, feces, mucus, or saliva. Deer are known to chew on the bones of dead deer, and this may contribute to transmission. It is also possible that CWD prions remain in the soil and water for years, thereby infecting healthy animals that are introduced into an area in which diseased herds have been eradicated. The practice of supplementary feeding of wild elk and deer, in order to increase their size and weight for hunting, may have contributed to the spread of CWD, as this concentrates the animals, enhancing contact.

As yet, there is no evidence that CWD crosses the species barrier into other domestic animals. CWD can be transmitted to cattle after intracerebral injection of CWD-infected brain tissue. However, it has not been observed to transmit to domestic cattle by ingestion of infected deer brain, or to cattle, sheep, or goats that have been in direct contact with CWD-infected deer or elk. However, further studies will be required. Because it is expected that CWD incubation time is prolonged in other species, it may take years for the disease to be detected in other species. Natural predators of deer and elk, such as cougars or coyotes, have not been observed to develop CWD. The disease has been transmitted by intracerebral injection into mice, ferrets, mink, goats, squirrel, and monkeys.

CAN HUMANS GET CWD?

It is frequently stated that there is no evidence that CWD can be transmitted to humans. However, as noted by the U.S. Centers for Disease Control, "there is not yet strong evidence that such transmissions could not occur." In recent years, several cases of CJD have occurred in people who were hunters and consumers of wild deer and elk in regions affected by CWD. These cases have been investigated, but no link to CWD has been established. They appear to have been cases of sporadic CJD. As the possible incubation time of CWD in nonruminant species is unknown, further monitoring will be required to firmly establish that a species barrier exists between deer and humans that would prevent transmission of CWD.

Also, because relatively few people would have been exposed to CWD (compared to those exposed to BSE in the United Kingdom), it may be difficult to detect the link. Another question is whether there are asymptomatic carriers of CWD that could pass the disease to others through blood transfusions or other iatrogenic routes such as organ transplants or contaminated surgical instruments. The incidence of CJD is similar in regions with and without CWD.

Despite the lack of evidence for transmission of CWD to humans, hunters have been cautioned about handling and eating venison. For example, the Wisconsin Game and Fish Department advises hunters to wear latex gloves when butchering deer and not to eat brain or spinal cord, eyes, spleen, or lymph nodes from deer or elk in affected areas of the country. There is concern that venison butchers may contaminate meat with neural tissue by using the same equipment to cut spinal cord and meat. As yet, there are no regulations governing venison processing by commercial or private butchers.

IS THERE AN EPIDEMIC OF CWD, AND IF SO, CAN IT BE STOPPED?

Both state and federal government agencies are seriously concerned about the rapid increase in CWD cases. In some regions, such as parts of northern Colorado, over 10% of wild deer and elk are infected. In other areas, about 1% test positive for CWD. In Wisconsin, about 3% of deer in south-central regions are affected. On one Nebraska domestic deer farm, 51% of animals were infected; wild deer surrounding the farm had a 3% to 7% incidence of CWD. Nebraska is concerned that the disease could spread rapidly among white-tail deer, threatening the state's entire deer herd. Computer models of CWD infection in Wisconsin predict that, if nothing is done to stop the spread of CWD, half the deer population will die within 25 years in the currently affected areas, and the disease will spread to the rest of the state.

Officials are concerned for several reasons. One is the obvious potential (but unproven) threat to human health. Another is the serious consequences to the economies of several states that depend upon income from hunting.

States and provinces vary considerably in their approaches to dealing with CWD. The practice of selling and transporting live deer and elk is now banned in most states. Wisconsin has one of the most aggressive control plans, authorizing the spending of $4 million in 2002 to 2003 to eradicate 25,000 deer within the affected regions in southern Wisconsin and to test up to 50,000 deer throughout the state. Two to three million pounds of deer carcasses will need to be incinerated during the cull program (Fig. 9). Wisconsin has also banned the

Figure 9
Wisconsin is conducting a major cull program to eradicate chronic wasting disease in deer.

practice of feeding and baiting wild deer. The U.S. Department of Agriculture has embarked on a program to test farmed elk populations, to eradicate herds containing CWD cases, and to compensate farmers. The effectiveness of these measures will take years to assess, owing to the lack of an effective screening test for CWD and the long incubation period between infection and appearance of symptoms. If it turns out that CWD prions persist in the environment for long periods and remain infective, eradication programs may only slow the progress of the disease, which may remain endemic in wild populations.

TSE Perspectives

Our understanding of TSEs is incomplete, uncertain, and often controversial. A great number of scientific questions about TSEs and their potential threats to human and animal health remain unanswered. How many people are infected with nvCJD but do not yet show symptoms? Is BSE-infected meat absent from the human food supply? Are TSEs spread through blood supply? Will CWD prove to be infectious to humans? Will cures be discovered for BSE, CWD, and the human TSEs?

In the face of such uncertainty, it has been difficult for governments to enact rational public policy decisions that would protect the maximum number of people, but not jeopardize the economic stability of important industries. The questions below illustrate a few of the difficulties involved in evaluating the risks of transmitting TSEs in order to protect both public and private interests.

HOW STRINGENT SHOULD WE MAKE GOVERNMENT REGULATIONS WHEN THE RISKS OF TRANSMITTING A TSE ARE UNKNOWN?

Early in the BSE epidemic, prior to 1996 when the first cases of nvCJD were documented, the British government repeatedly emphasized the low risks that BSE posed to humans, and encouraged the consumption of beef (see Fig. 10). In hindsight, this was a serious mistake, but at the time little was known about transmissibility of TSEs between species, and even less about BSE, which represents a special case of interspecies transmission. It was assumed that BSE had

Figure 10
John Gummer, Agriculture Minister in the UK Conservative government, posing for the cameras with his daughter Cordelia and a couple of burgers—May 1990. "My wife eats beef, my childern eat beef, and I eat beef," he said. "That is everyone's absolute protection." The British government's attempts to reassure the public, even after it had offically banned beef for human comsumption, contributed to the public's mistrust of government pronouncements on food safety.

arisen from scrapie in sheep, and therefore would be as benign as scrapie is to humans. The British government was reluctant to frighten people unnecessarily, or to cause a panic that would disrupt the $6 billion per year beef industry. Many scientists, in contrast, stressed that the risks were unknown, and that policies and regulations should be based on the assumption that risk is high until proven otherwise.

This question is equally relevant today when thinking about the CWD epidemic in North America. Should government policies be based on the assumption that CWD poses a serious health threat to humans, even though there are no known cases of human CWD? Or should we assume low risk, until a case of human CWD is detected? It might be argued that the potential scale of the problem is smaller with CWD than with the BSE epidemic, given the fact that deer and elk meat are not major components of human food in North America. However, the questions of ethics, unknown risks, and public policy are the same as those involved with the British Mad Cow Disease epidemic.

SHOULD PEOPLE AT RISK FOR CONTRACTING nvCJD BE INFORMED WHEN THE RISK OF INFECTION IS UNKNOWN AND THE DISEASE IS INCURABLE?

In general, ethicists agree that full disclosure of risks is necessary in cases where the public or individuals have been placed at unknown or uncertain risk of contracting a serious disease such as nvCJD. Not everyone agrees with the decision to inform, though, especially when there is little immediate benefit to the patients themselves, and perhaps extra difficulties, such as unnecessary stress or barriers to obtaining insurance. The following examples illustrate these viewpoints.

The United Kingdom is conducting a survey of human tonsil and appendix samples to estimate the number of asymptomatic patients that may be incubating nvCJD. The study will be anonymous, meaning that patients cannot be informed of the results. In contrast, a similar study in Switzerland will retain the patient's identities and will inform those who request the information, if a therapy becomes available at a later date.

In Canada, 71 people who had undergone endoscopy at a Saskatoon hospital were informed that the endoscope that was used for their procedure had been used previously on a man who was later diagnosed with nvCJD. Although the risk of prions being transferred from the equipment after multiple cleanings and sterilizations was considered to be extremely low, hospital officials felt obliged to inform the patients of their risk.

In Britain, 22 patients received transfusions of blood that was later discovered to have been donated by nvCJD patients. The risks of transmitting nvCJD through blood are still unknown, although animal studies suggest the possibility of such transmission. The UK blood services first decided not to inform the recipients, as there is no treatment or cure for the disease and the risks of infection from blood are unknown. It recently changed this policy and decided to inform the recipients, so that they could refrain from making blood donations or organ donations in the future and take precautions when they undergo surgical or dental procedures.

COULD PRIONS OF AN ENTIRELY NEW TSE ENTER OUR FOOD SUPPLY, CAUSING DISEASE IN HUMANS?

The BSE epidemic provides a chilling example of how humans change their environments in ways that encourage the emergence of new diseases. In the 1940s, Western countries began feeding protein supplements to farm animals to increase growth, production, and profits. These supplements were prepared from the rendered remains of other farm animals — a kind of forced cannibalism. The feeding of sheep and cattle to other sheep and cattle, combined with changes in rendering practices, spread BSE rapidly, turning it into an epidemic. This herbivore cannibalism may also have recycled the disease within cattle populations, leading to creation of new strains of prions never seen before — strains that could leap the species barrier into cats, zoo animals, and humans.

In the future, it is possible that other food-borne TSE diseases could emerge and spread as a result of modern agricultural practices. At present, animal feed regulations in North America still permit the feeding of rendered pork, chicken, and fish to other animals, and rendered ruminants to nonruminants. These measures may or may not be sufficient to prevent a newly emerged TSE from cycling within domesticated animals and leaping the species barrier into humans. The lessons learned from Mad Cow Disease and kuru in humans should encourage us to assess carefully our agricultural practices and demand stringent monitoring of our food supply.

References and Resources

Scientific Papers

TSEs and Prions

Aguzzi, A. and Weissman, C. Prion research: the next frontiers. Nature 389: 795–798 (1997).

Balter, M. On the hunt for a wolf in sheep's clothing. Science 287: 1906–1908 (2000).

Belay, E.D. Transmissible spongiform encephalopathies. Annu. Rev. Microbiol. 53: 283–314 (1999).

Bosque, P.J. Prions in skeletal muscle. Proc. Natl. Acad. Sci. USA 99: 3812–3817 (2002).

Brown, P. Drug therapy in human and experimental transmissible spongiform encephalopathy. Neurology 58: 1720–1725 (2002).

Collinge, J. Prion diseases of humans and animals: their causes and molecular basis. Annu. Rev. Neurosci. 24: 519–550 (2001).

Hunter, N. et al. Transmission of prion diseases by blood transfusion. J. Gen. Virol. 83: 2897–2905 (2002).

Prusiner, S.B. Prions. (Nobel Lecture). Proc. Natl. Acad. Sci. USA 95: 13363–13383 (1998).

Van Everbroeck, B. et al. Transmissible spongiform encephalopathies: the story of a pathogenic protein. Peptides 23: 1351–1359 (2002).

Weber, D.J. and Rutala, W.A. Managing the risk of nosocomial transmission of prion diseases. Curr. Opin. Infect. Dis. 15: 421–425 (2002).

BSE and nvCJD

Balter, M. Tracking the human fallout from 'Mad Cow Disease'. Science 289: 1452–1454. (2000).

Brown, P. et al. Bovine spongiform encephalopathy and variant Creutzfeldt-Jakob disease: background, evolution and current concerns. Emerg. and Infect. Dis. Vol. 7 (Jan/Feb 2001). Published on: **http://www.cdc.gov/ncidod/EID/vol7no1/brown.htm**

Bruce, M.E. et al. Transmissions to mice indicate that "new variant" CJD is caused by the BSE agent. Nature 389: 498–501 (1997).

Collee, J.G. and Bradley, R. BSE: a decade on - part 1. The Lancet 349: 636–641 (1997).

Collee, J.G. and Bradley, R. BSE: a decade on - part 2. The Lancet 349: 715–721 (1997).

Collinge, J. et al. Molecular analysis of prion strain variation and the aetiology of new variant CJD. Nature 383: 685–690 (1996).

Enserink, M. Is the U.S. doing enough to prevent Mad Cow Disease? Science 292: 1639–1641 (2001).

Ferguson, N.M. Estimating the human health risk from possible BSE infection of the British sheep flock. Nature 415: 420–424 (2002).

Foster, J.D. et al. Distribution of the prion protein in sheep terminally affected with BSE following experimental oral transmission. J. Gen. Virol. 82: 2319–2326 (2001).

Hill, A.F. et al. The same prion strain causes nCJD and BSE. Nature 389: 448–450 (1997).

Lasmezas, C.I. et al. BSE transmission to macaques. Nature 381: 743–744 (1996).

O'Brien, M. Have lessons been learned from the UK bovine spongiform encephalopathy (BSE) epidemic? Int. J. Epidemiol. 29: 730–733 (2000).

Prusiner, S. B. Prion diseases and the BSE crisis. Science 278: 245–251 (1997).

Wadman, M. Agencies face uphill battle to keep United States free of BSE. Nature 409: 441–442 (2001).

Will, R.G. et al. A new variant of Creutzfeldt-Jakob disease in the UK. The Lancet 347: 921–925 (1996).

Kuru

Gajdusek, D.C. Unconventional viruses and the origin and disappearance of kuru. Nobel Lecture, December 13, 1976. Published on: **http://www.nobel.se/medicine/laureates/1976/gajdusek-lecture.html**

CWD

Knight, J. Ranches blamed over spread of mad deer. Nature 416: 569–570 (2002).
Dalton, R. and Check, E. Prion research stepped up as fear grows of deer disease. Nature 419: 236 (2002).

WEB SITES

http://www.bsereview.org.uk
The UK Food Standards Agency review of BSE regulations, information on BSE, latest news and statistics on BSE.

http://www.cjd.ed.ac.uk
The UK Creutzfeldt-Jakob Disease Surveillance Unit, University of Edinburgh statistics and background information on CJD and nvCJD.

http://www.bseinquiry.gov/uk/index.htm
Report of the UK public inquiry into the BSE epidemic, October 2000. Contains background information, findings, and conclusions from the inquiry.

http://www.ifst.org/hottop5.htm
The UK Institute of Food Science & Technology Web site, with information and statistics on BSE and nvCJD from the United Kingdom and around the world.

http://www.newscientist.com/hottopics/bse/
Links to articles in the New Scientist, both news items and background on BSE and nvCJD.

http://www.priondata.org
Extensive resource for news on BSE, nvCJD, and other TSEs.

http://www.maddeer.org
Links to news items, background information, and research on CWD.

http://www.organicconsumers.org/madcow.htm
Links to news articles on BSE and CWD.

http://www.cdc.gov/ncidod/diseases/cjd/bse_cjd.htm
U.S. Centers for Disease Control Web site. Information on BSE and CJD.

http://www.cdc.gov/ncidod/diseases/cjd/cjd_fact_sheet.htm
U.S. Centers for Disease Control Web site. Information on nvCJD.

http://www.fda.gov/oc/opacom/hottopics/bse.html
The U.S. Food and Drug Administration Web site, with links to BSE and CJD information. Also, blood safety, ruminant feed regulations, and traveler's information.

http://www.aphis.usda.gov/lpa/issues/cwd/cwd.html
The U.S. Animal and Plant Health Inspection Service (USDA) Web site, with links to CWD information and regulations.

Other Media

http://www.guardian.co.uk/bse

News coverage and links to sources of information about the BSE epidemic. Extensive archives of BSE-related news stories.

http://www.pbs.org/wgbh/nova/madcow/

Transcript and resources to accompany the NOVA program, "The Brain Eater," February 10, 1998.